Developing Operationally Relevant Metrics for Measuring and Tracking Readiness in the U.S. Air Force

MUHARREM MANE, ANTHONY D. ROSELLO, PAUL EMSLIE,
THOMAS EDWARD GOODE, HENRY HARGROVE, TUCKER REESE

Prepared for the Department of the Air Force
Approved for public release; distribution unlimited

For more information on this publication, visit www.rand.org/t/RRA315-1

Library of Congress Cataloging-in-Publication Data is available for this publication.
ISBN: 978-1-9774-0609-5

Published by the RAND Corporation, Santa Monica, Calif.
© Copyright 2020 RAND Corporation
RAND® is a registered trademark.

Cover: Staff Sgt. Justin Parsons/U.S. Air Force; Sammby/Adobe Stock.

Support RAND

Make a tax-deductible charitable contribution at
www.rand.org/giving/contribute

www.rand.org

Preface

In 2019, Ted Uchida (deputy director of operations, Air Combat Command [ACC]) commissioned RAND Project AIR FORCE to examine readiness data across ACC and identify a methodology for aggregating the data to enable the measurement of readiness in terms of combat power available to commanders. The research was conducted within the Force Modernization and Employment Program of RAND Project AIR FORCE as part of a fiscal years 2019 and 2020 project. The project built on work conducted between February 2018 and January 2019 that analyzed and assessed the current readiness reporting approach and data. We examined how authoritative data sources that feed the current readiness reporting system can be leveraged and aggregated to produce more-relevant readiness metrics and to help decisionmakers gain a better picture of the actual combat power capability available at a given time across ACC wings. This report describes the methodology, analysis, and findings of this project. It should be of interest to those interested in military readiness reporting and readiness improvement.

RAND Project AIR FORCE

RAND Project AIR FORCE (PAF), a division of the RAND Corporation, is the Department of the Air Force's (DAF's) federally funded research and development center for studies and analyses. PAF provides the DAF with independent analyses of policy alternatives affecting the development, employment, combat readiness, and support of current and future air, space, and cyber forces. Research is conducted in four programs: Strategy and Doctrine; Force Modernization and Employment; Manpower, Personnel, and Training; and Resource Management. The research reported here was prepared under contract FA7014-16-D-1000.

Additional information about PAF is available on our website:
www.rand.org/paf/

This report documents work originally shared with the DAF on April 27, 2020. The draft report, issued on May 5, 2020, was reviewed by formal peer reviewers and DAF subject-matter experts.

Contents

Figures

Tables

Summary

Issue and Approach

Current force readiness and availability metrics have important deficits that limit their ability to inform U.S. Air Force decisionmakers about the number of units available and to identify capability and capacity shortfalls in meeting scenario demands. For this project, we developed an approach that allows authoritative data sources that feed the current readiness reporting system to be leveraged and aggregated and therefore better measure the readiness of Air Combat Command forces to meet scenario demands. The methodology proposes the definition of *combat power* as the specific collection of personnel and equipment to fulfill a given capability (e.g., the air superiority capability of a six-ship of F-22 aircraft). Combat power readiness would be measured by linking the status of personnel and equipment to specific capability sets (e.g., fly the aircraft, maintain the aircraft) that contribute to the required set of Unit Type Codes (UTCs) demanded by scenarios when said power would be utilized.

Findings

- The existing UTC construct provides a foundation for defining *combat power* and measuring its readiness.
- The proposed approach can be used to assess combat power readiness from a manpower perspective. Shortfalls in personnel availability tend to drive low readiness; thus, a personnel assessment provides a close approximation of reality.
- Improving access to some existing data and information would enable combat power readiness to be assessed and would provide decisionmakers with a metric that informs the availability of usable capability.

Recommendations

- Track and make available personnel data at the individual level and equipment data at the part level. Including more-detailed personnel and equipment data in existing systems (e.g., Defense Readiness Reporting System–Strategic) will help to complete the readiness assessment.
- Explicitly link personnel availability data to personnel training data. Personnel availability and training data must be considered and accessed simultaneously at the individual level, since bringing together availability and training data at the unit level overlooks important information.
- Define or update descriptions and groupings of UTCs that enable their aggregation into relevant force packages. A more clear and concise description of mission capability statements of the UTCs is needed to provide the foundation for the grouping of UTCs into combat power.

Acknowledgments

We would like to thank the project sponsor, Ted Uchida, Air Combat Command, A3-2; his dedicated team, including our action officer, Joseph Morina; Col John Lussier; and Steven Scroggins. We would also like to thank our RAND colleagues Daniel Norton and Jeffrey Kendall, who improved the report through their constructive reviews of an earlier version.

Abbreviations

ABW	air base wing
ACC	Air Combat Command
AETF	air expeditionary task force
AFB	Air Force base
AFI	Air Force instruction
AFSC	Air Force Specialty Code
ANG	Air National Guard
ART	Air Expeditionary Force (AEF) Reporting Tool
C2	command and control
DoD	U.S. Department of Defense
DRRS-S	Defense Readiness Reporting System–Strategic
FP	force package
FW	fighter wing
MEFPAK	manpower and equipment force package
MET	mission-essential task
MilPDS	Military Personnel Data System
MISCAP	Mission Capability Statement
NDS	National Defense Strategy
NSN	National Stock Number
OPLAN	operation plan
UTC	Unit Type Code

1. Introduction

Background

The unclassified *Summary of the 2018 National Defense Strategy* (referred to here as the NDS 2018) identifies long-term strategic competition with near-peer nations as a principal priority for the U.S. Department of Defense (DoD) and reiterates that DoD's enduring mission is to provide combat-credible military forces to deter war—and, should deterrence fail, to win the war.[1] As stated in NDS 2018, prioritizing preparedness for war means that "during normal day-to-day operations, the Joint Force will sustainably compete to deter aggression, degrade terrorist and WMD [weapon of mass destruction] threats, and defend U.S. interests from challenges below the level of armed conflict." Similarly, NDS 2018 states that, in "wartime, the fully mobilized Joint Force will be capable of defeating aggression by a major power, deterring opportunistic aggression elsewhere, and disrupting imminent terrorist and WMD threats." These two strategic missions translate into two types of scenarios for which U.S. forces must be ready and able to respond: competition and wartime, each with their different priorities.

Having an operationally relevant readiness metric that supports deliberate planning, contingency planning, execution, and resource-allocation decisions to improve readiness for these scenarios is essential. Specifically, this metric is needed to inform decisionmakers of the force's ability to perform its missions and identify shortfalls in—as well as their impact on—the combat power capabilities required in each scenario. By *combat power*, we are referring to the specific collection of personnel and equipment to fully provide a given capability—for example, the capability to operate a six-ship of F-22s.[2] Combat power is not simply the operations squadron to which the pilots are assigned. Nor is it solely the maintenance squadron that reports on the status of aircraft and that can generate sorties but not fly them.[3] In this report, and in this F-22 example, we define *combat power* as all the personnel and equipment across the operations, maintenance, and munitions squadrons required to employ a six-ship of F-22s.

Combat power is a somewhat nonstandard phrase, but it has been chosen carefully in this context and for this report to avoid confusion with other terms and concepts. Some terms and

---

[1] DoD, *Summary of the 2018 National Defense Strategy of the United States of America: Sharpening the American Military's Competitive Edge*, Washington D.C., January 2018.

[2] The Unit Type Code (UTC) source documents, the manpower and equipment force package (MEFPAK) and Mission Capability Statement (MISCAP), refer to different aviation packages using the *X-ship* language, which we maintain for consistency. In other usages, the *X-ship* language refers to the actual number of aircraft flying in a formation rather than the number of aircraft that are expeditionarily employed.

[3] In U.S. Air Force readiness reporting, aircraft are considered to be "equipment" assigned to maintenance squadrons that report on the status of those aircraft. In other Air Force contexts, aircraft are considered as belonging to the operations squadrons.

phrases were rejected to avoid confusion and to differentiate from similar terms used in the existing readiness reporting systems and processes of the Air Force. For example, in DoD's Defense Readiness Reporting System–Strategic (DRRS-S), the system of record for DoD readiness reporting, *capability readiness*, *availability*, and *unit* have specific connotations that are different from our more general intent here with *combat power*. Additionally, we stayed away from terms related to *deployability* to avoid confusion with issues related directly to the transport of expeditionary units to a place of operations. We avoided *unit availability* because the combined sets we refer to later come from below the squadron level but are gathered from multiple squadrons. In the world of Air Force readiness reporting, *unit* almost exclusively refers to individual squadrons, independently. Similarly, *unit availability* would be problematic because the phrase already has specific meaning within the readiness reporting system related to personnel who are both assigned and able to deploy. Thus, although *combat power* is somewhat nonstandard, we continue with the phrase in this report to avoid confusion with these other terms and phrases that presently carry specific meanings.[4]

Given this definition of *combat power*, a methodology that leads to the readiness assessment of combat power should be able to

- quantify combat power to give a sense of capacity
- identify the contributors to that combat power
- link each contributor to the readiness of that combat power
- assess the readiness of combat power as a function of the conditions and requirements of each scenario.

The existing foundation for defining and quantifying combat power is the UTC. As described in AFI 10-401,[5] a UTC is "a potential capability focused on accomplishment of a specific mission that the military Service provides." It is composed of personnel or equipment with the necessary skills and features, respectively, to accomplish the assigned mission.[6] The MISCAP of each UTC, which defines the basic mission for which the UTC is designed, is used by war planners to determine which UTCs can fill their requirements.

Although UTCs describe a capability to accomplish a mission, a combination of UTCs is what defines *combat power*. Keeping with the same F-22 example above, two UTCs describe the maintenance capability to support a six-ship of F-22s (one for maintenance personnel and one for

---

[4] For more details on the usage of these terms, see Air Force Instruction (AFI) 10-201, *Force Readiness Reporting*, Washington, D.C.: U.S. Air Force, March 3, 2016.

[5] AFI 10-401, *Air Force Operations Planning and Execution*, Washington, D.C.: U.S. Air Force, December 7, 2006, p. 87.

[6] In other documents and contexts, UTCs are not described simply as capabilities in and of themselves but rather descriptors of a unit's capability based on the force type. For example, the UTC "is associated with and allows each type organization to be categorized into a kind or class with common distinguishing characteristics. The UTC is one of the primary means for identifying types of forces when describing force requirements" (Chairman of the Joint Chiefs of Staff Manual 3150.24E, *Type Unit Characteristics Report (TUCHAREP)*, Washington, D.C.: Joint Chiefs of Staff, August 10, 2018, Appendix A, p. A-A-1).

data-availability shortfalls that impede such a presentation. The final chapter summarizes the findings and presents recommendations.

Appendix A presents a more detailed description of current readiness metrics. Appendix B provides a description of UTCs from MISCAPs and the MEFRK.

2. Current Readiness Reporting and Its Shortfalls

In this chapter, we examine how the current readiness reporting system works to capture and generate readiness information that allows commanders to readily assess their combat capability. Then, we look at some of the shortfalls of that system.

Current Readiness Reporting System

Currently, the primary source for readiness tracking and readiness data is the Defense Readiness Reporting System–Strategic (DRRS-S), the official system the Air Force uses to report resource and capability readiness to DoD. The Air Expeditionary Force (AEF) Reporting Tool (ART) is used by the Air Force to track unit readiness.[8]

As prescribed in AFI 10-201, units report in DRRS-S on both resource and unit readiness. A readiness reporting unit is an individual squadron. The readiness information reflects the readiness of the unit as whole. However, as described in AFI 10-244, ART provides one central location to archive unit commander assessments on the status of individual UTC's. However, the information cannot be readily aggregated to provide measures of the combat capability that a given set of UTCs can provide. In fact, AFI 10-244 states that "ART is not a report card, but a method of identifying a UTC's ability to perform its MISCAP and identify shortages in resources." In addition, AFI 10-401, para. 4.4.6, states "ART does not measure a UTC's availability to deploy, only its ability to meet its MISCAP should it be tasked."[9] Therefore, ART data must still be arranged and aggregated to provide information about unit capabilities.

In DRRS-S, *resource readiness* (formerly known as the Status of Resources and Training System [SORTS]) represents an objective assessment of the unit's ability to execute the full range of missions for which the unit is organized. In addition, resource readiness measures the Air Force's effectiveness in meeting its responsibilities to organize, train, and equip forces for combatant commands. A related, but independent, readiness measurement within DRRS-S, *capability readiness*, is the commander's subjective assessment of the unit's ability to conduct discrete mission-essential tasks (METs) with the trained personnel and equipment assigned to it. Appendix A presents a more detailed description of these metrics.

These three assessments (DRRS-S resource, DRRS-S capability, and ART), while related, are distinct; use different measures, criteria, and reporting rules to express readiness; and

---

[8] AFI 10-244, *Reporting Status of Air and Space Expeditionary Forces*, Washington, D.C.: U.S. Air Force, June 15, 2012.

[9] AFI 10-244, 2012.

together are meant to provide an assessment of a unit's readiness to perform its mission (see Figure 2.1).

Figure 2.1. Overview of Contributing Components Readiness Assessment in DRRS-S

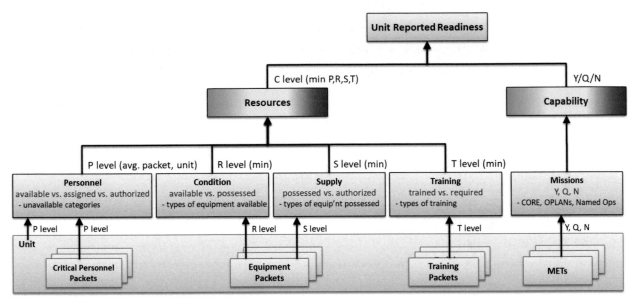

SOURCE: RAND analysis of DRRS-S information.
NOTE: C = resource readiness level; min = minimum; P = personnel; R = equipment condition; S = supplies on hand; T = training of personnel; Y = yes; Q = qualified yes; N = no; CORE = Core Mission Essential Task List; OPLAN = operation plan; ops = operations.

At its core, the resource readiness metric aggregates information about personnel availability, personnel training, and equipment supply and condition of the reporting unit, without considering when and how these personnel and equipment would be used. Capability readiness, however, attempts to capture the ability of a unit to perform its assigned missions but does not consider dependencies among units or measure combat power capability. The type of readiness information available from DRRS-S is summarized in Figure 2.2 through a notional example for the F-22 operation and maintenance units at the 1st Fighter Wing (FW) and 192nd FW at Langley Air Force Base (AFB) in Virginia.

Figure 2.2. Current Readiness Metrics Generated by DRRS-S for a Notional Example at Langley AFB

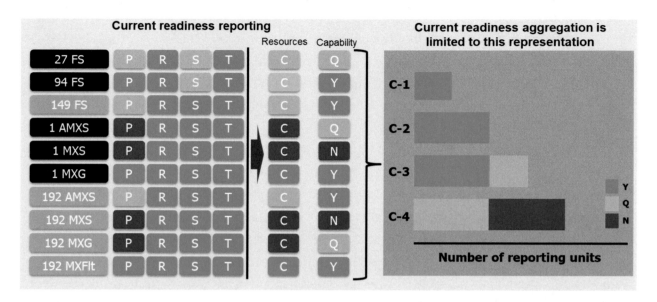

NOTE: FS = fighter squadron; AMXS = aircraft maintenance squadron; MXS = maintenance squadron; MXG = maintenance group; MXFlt = maintenance flight; P = personnel; R = equipment condition; S = supplies on hand; T = training of personnel; C = resource readiness level; Y = yes; Q = qualified yes; N = no.

The personnel, equipment condition, equipment supply, and personnel training status of each unit determines the resource readiness level, or C-rating, of the unit, and the commander's assessment of the unit's capability determines the capability readiness.[10] Taken together, it is possible to build an overview of the fraction of the units that are at acceptable readiness levels and those that are not (see Figure 2.2, right side).

Shortfalls of Current Readiness Reporting

Although this approach is useful in identifying individual units that need improvement, the current view of readiness does not provide enough information to determine the amount of combat power that units can generate. DRRS-S focuses on squadrons as the smallest measured

---

[10] The DRRS-S resource areas of P, R, S, and T and the capability assessments of Y, Q, and N are described in more detail in AFI 10-201, 2016. Briefly, P = personnel; S = supplies on hand; R= equipment condition; and T = training of personnel. P and T relate to a unit's people: P captures the number available compared with authorized, and T captures the proportion of the unit's assigned personnel who are properly trained. S and R are the equipment analogues. S captures the equipment that a unit possesses compared with what it is supposed to have. R captures whether the equipment the unit possesses is functioning properly. The DRRS-S capability assessments of Y, Q, and N (yes, qualified yes, and no, respectively) describe the commanders' assessments of whether their units can execute their assigned mission and tasks.

entity.[11] It does not consider the UTCs that are needed to generate specific capabilities. Additionally, units are stand-alone entities in the DRRS-S system, whereas, in reality, units are interdependent.

The ART approach goes below the unit level and focuses on UTC capabilities, but it does so by considering units in isolation. In fact, multiple units in the same wing could contribute personnel and equipment to fill UTCs. Furthermore, ART does not aggregate unit-specific capabilities into a deployable combat power; a wing-level readiness assessment of capability is not the norm.

Consider a deployable F-22 squadron. An aviation package (operations, maintenance, and munitions capabilities) of F-22s can be deployed as a six-ship lead package with up to three follow-on elements (follow 1, follow 2, and follow 3), as shown in Figure 2.3.[12] Each element is composed of six UTCs: operations, maintenance, and munition personnel and operations, maintenance, and munition equipment. Each UTC represents a capability that a unit (or multiple units) must generate whose readiness is measured and tracked. For example, 3FBP1, HFBP1, and HGBP1 are the personnel UTCs for operations, maintenance, and munitions, respectively, for a six-ship lead package, and 3FBE1, HFBE1, and HGBE1 are the equipment UTCs for operations, maintenance, and munitions, respectively, for a six-ship lead package. UTCs 3FBP1, HFBP1, and HGBP1 are composed of 92, 316, and 90 airmen, respectively, and of 11, 35, and three different AFSCs each.[13]

Figure 2.3. UTCs for Generating Deployable F-22 Aviation Packages

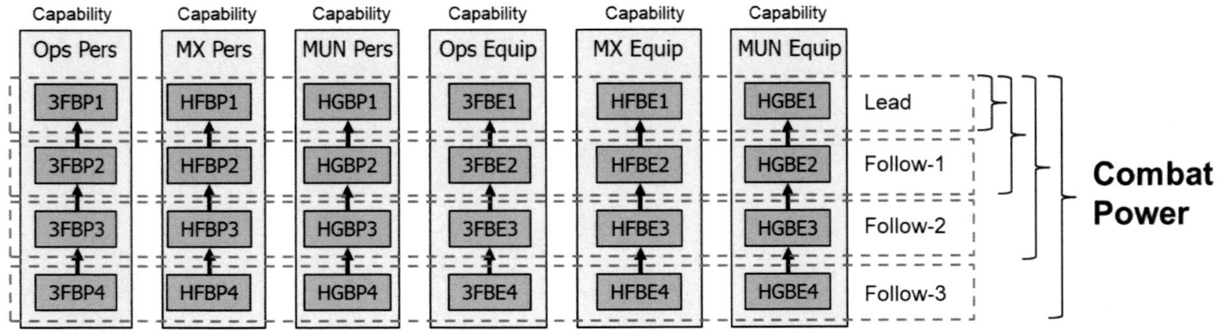

NOTE: Ops = operations; Pers = personnel; MX = maintenance; MUN = munitions.

---

[11] Most units do report on the resource and capability of critical personnel and critical equipment packets; these are subsets of people and equipment within a unit that are deemed critical for completing the unit's mission. However, the definition of *critical packets* is mostly made along functional lines or skills and not from the perspective of capability generation. For example, a personnel critical packet can be composed of all the personnel of a specific skill level for a specific Air Force Specialty Code (AFSC). See AFI 10-201, 2016.

[12] According to AFI 10-401, 2006, "Lead packages will be designed to deploy independently. Follow-on UTCs will be dependent packages that fall in on (provide additional aircraft capabilities to) initial lead packages."

[13] Data are from Air Force's Manpower and Equipment Force Packaging System.

Current readiness reporting focuses on the readiness of each UTC (i.e., each capability) and does not aggregate that readiness to account for how UTCs are actually employed together. All six UTCs must be available (manned and equipped) for the six-ship air superiority package (combat power) to be deployable. They compose the minimum requirement for the "construction" of the air superiority capability. The availability of personnel and equipment of any of these UTCs in isolation does not provide combat power. Furthermore, the availability of any of the follow-on packages in isolation is irrelevant if the lead package is not already available or deployed.

In addition, the actual personnel and equipment to fill these UTCs come from different units (squadrons); in some instances, as is the case for the F-22s at Langley AFB, they come from a combination of the active duty, Air National Guard (ANG), and reserve component of Air Force (Figure 2.4).

Figure 2.4. UTC, Squadron, and Air Force Component Interdependencies in Deployable F-22 Aviation Packages at Langley AFB

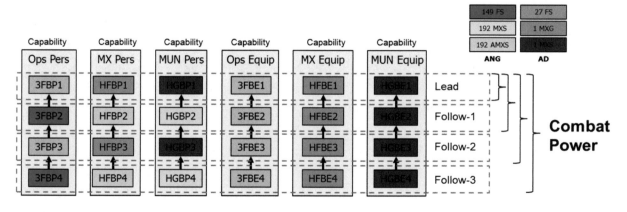

NOTE: Ops = operations; Pers = personnel; MX = maintenance; MUN = munitions; Equip = equipment; FS = fighter squadron; MXS = maintenance squadron; MXG = maintenance group; AMXS = aircraft maintenance squadron; MXS = maintenance squadron; AD = active duty.

For instance, assessing the ability of Langley AFB to generate a deployable six-ship lead air superiority aviation package requires the simultaneous consideration of three active duty squadrons (27th Fighter Squadron, 1st Maintenance Group, and 1st Maintenance Squadron). Similarly, to generate a full squadron of aviation combat power with air superiority, the readiness of six squadrons across the active duty and ANG components must be considered simultaneously.

Current measurement and tracking of the resource readiness of each squadron provides information about the availability of personnel and equipment of the individual squadron.

Although this is useful information to identify shortcomings of individual whole squadrons, this falls short in informing decisionmakers about whether the squadron can generate any minimum amount of combat power, because one must look across squadrons to make this determination. Additionally, the current resource readiness approach does not account for how combining personnel from across multiple similar units could generate complete UTCs. Capability readiness for maintenance units gets close to providing this information, but it is still lacking. By capturing the commanders' assessment of the METs' readiness, a unit reports on its own ability to generate the UTCs for which it is responsible, but there is no aggregation of this information above the squadron level across multiple units to allow, say, the wing commander to report on the ability of the wing to generate combat power.

Similar limitations exist in this representation of this readiness information. Readiness levels of individual UTCs are tracked and reported by each unit responsible for them, but they are not aggregated across units to report the readiness of the combat power that groups of UTCs provide.

Summary

This chapter presented a short overview of the existing readiness reporting processes and system (DRRS-S). A major shortfall of DRRS-S is that it focuses on units independently and thus does not allow an assessment of whether all the contributing parts below the unit level (which must be combined to provide combat power) are ready at the same time. UTCs—the collections of personnel and equipment at the subunit level—were introduced, along with the concept that UTCs are combined to produce elements of combat power.

3. An Approach to Measure Operationally Relevant Readiness and the Challenges to Implementing It

In this chapter, we discuss our proposed approach to measure operationally relevant readiness that addresses the shortfall of the current readiness system, discussed in the previous chapter. Here, we describe the approach and then review the challenges of implementing it.

An Approach to Measure Readiness in an Operationally Relevant Manner

Combat air forces fighter and bomber squadrons are aligned into five general mission areas: (1) air superiority, (2) global precision attack, (3) suppression of enemy air defenses, (4) close air support, and (5) long-range strike capabilities. The readiness assessment for any of those mission areas is a function of scenario demands. For example, a wartime scenario, as articulated in the NDS,[14] involving a nearly all-in effort of entire fighter squadrons would lead to one assessment. A smaller rotational presence of lead packages supporting ongoing "competition" missions would lead to another. Similarly, deployments to different geographical locations with varying levels of base support, against a different enemy, and within different timelines would have different requirements and therefore lead to different readiness assessments.

In fact, a detailed description of these conditions is needed to provide an effective readiness measure. In Table 3.1, we propose six broad categories that can be used to describe a scenario for which readiness can be assessed: threat, response time, response intensity, conflict duration, command and control (C2) structure, and air base support. Additional categories could be added if the capabilities associated with each can be quantified.

For the six categories and the four types described in Table 3.1, there are more than 3,072 possible scenarios that span the space, from "clear-cut" to "highly complex" from a readiness perspective.[15] For example, consider the notional scenarios presented in Table 3.2.

---

[14] DoD, 2018.

[15] The 3,072 possible combinations stem from five of the six categories having four options and the sixth category having three options. $3,072 = 4 \times 4 \times 4 \times 4 \times 4 \times 3$.

Table 3.1. Notional Description of Dimensions That Can Be Used to Create Readiness Assessment Scenarios

Category	Description	Type 1	Type 2	Type 3	Type 4
Threat	Type of threat	Level 1	Level 2	Level 3	Level 4
Response time	Amount of time to constitute response force	24 hours	72 hours	1 week	1 month
Response intensity	Type of package deployed	Lead	Lead + Follow 1	Lead + Follow 1 + Follow 2	Full squadron
Conflict duration	Duration of deployment	0–7 days	7–30 days	30–180 days	180+ days
C2 structure	Type of C2 structure required	Air expeditionary squadron	Air expeditionary group	Air expeditionary wing	Numbered expeditionary air forces
Air base support	Type of air base support needed	Bare base	Forward operating base	Main operating base	Not applicable

Table 3.2. Notional Deployment Scenarios for Which Readiness Can Be Assessed

Category	Scenario 1 (clear- cut)	Scenario 2	. . .	Scenario N (highly complex)
Threat	Level 1	Level 2	. . .	Level 4
Response time	1 month	72 hours	. . .	24 hours
Response intensity	Lead element	Full squadron	. . .	Full squadron
Conflict duration	0–7 days	30+ days	. . .	180+ days
C2 structure	Air expeditionary squadron	Air expeditionary squadron	. . .	Numbered expeditionary air forces
Air base support	Main operating base	Bare base	. . .	Bare base

Scenario 1 is the most benign and has the fewest requirements. It calls for the ability to respond to a low-level threat, has a preparation time of one month, requires the deployment of only a lead element for up to seven days to a main operating base, and uses only an air expeditionary squadron C2 component. Therefore, readiness for such a scenario should be the easiest to achieve. In contrast, the 180-day deployment of a full squadron against a high-level threat, within 24 hours, to a bare base, and requiring a C2 component from one of the numbered expeditionary air forces, can be a very demanding scenario.

Specific combinations of UTCs are needed to meet capability requirements. The set of UTCs needed in scenario 1 is different from those required to respond to scenario 2 (or any other scenario). For example, the threat level may require different training proficiencies, the response time may require the use of personnel and equipment that can be available within 24 hours, a response that requires the deployment of a lead package will need fewer (and different) UTCs than a response that requires the deployment of a full squadron, deployment to a main operating base will have different requirements from deployment to a bare base, and so on.

The information and data to enable the readiness assessment of a wing (or combination of wings) to generate the combat power required for each scenario must enable the grouping, aggregation, and description of UTCs for each scenario category. Thus, a readiness assessment of the capabilities needed requires

- information about the supply of personnel and equipment—i.e., information and data about the training and availability of personnel and the availability of equipment in each unit
- a description of the personnel and equipment in each UTC
- the listing and description of the combination of UTCs needed to provide a given capability
- a description of the capabilities needed for a given scenario
- an analytical model that matches the available personnel and equipment to fill the needed UTCs to provide the given capabilities needed for each scenario.

These components are described in the following sections.

Personnel, Training, and Equipment Data

Assessing the readiness of units to provide needed capabilities requires aggregating information about personnel, training, and equipment in an appropriate and meaningful way. Necessary personnel information includes the AFSC of an individual and that person's availability status. The National Stock Number (NSN) of an individual piece of equipment is needed to assess its availability. Personnel training status is based on specialty, rank, and qualifications. Taken together, this information composes the pool of resources available to a squadron and wing.

The systems and databases used by the Air Force to record and track this information are numerous and not always interconnected. Personnel data that describe the status and availability of each airman reside in the Air Force's Military Personnel Data System (MilPDS). Data that describe and track the training requirements and status of airmen reside in different databases for different communities (e.g., Training Business Area for enlisted personnel, Automated Aircrew Management System for aircrews, and Automated Civil Engineer System for civil engineers). Equipment data are also in numerous databases, although the recently released Defense Property Accountability System promises to be a one-stop shop for equipment-related information.

UTC Data

UTCs use their assigned personnel and equipment to provide capabilities. In this section, and through the rest of the report, we use the example of F-22s and the units at Langley AFB. The 1 FW, 192 FW, and 633rd Air Base Wing (ABW) at Langley AFB enable the deployment and operation of F-22 squadrons and are responsible for 342 UTCs in 26 units, with more than 4,000 authorized personnel. However, some UTCs overlap in the capabilities they provide.

Table 3.3 gives an example of the aviation UTCs (operations, maintenance, and munitions) for which the 1 FW and 192 FW are responsible. The aviation UTCs are, in fact, some of the better-defined UTCs, whose capabilities are clear. (See Appendix B for a look at issues in the description of UTCs and their capabilities.)

Table 3.3. Select UTCs from 1 FW and 192 FW at Langley AFB

Wing	UTC Type	UTC	Mission Capability Description	Number of Distinct Units	Number of Authorized Personnel
1 FW	Operations	3FBP1	Provides operations personnel to support a six-ship lead package for contingencies or general war	2	92
		3FBP3	Provides additional operations personnel to support a follow on six-ship package for contingencies or general war	1	28
	Maintenance	HFBP1	Provides an independent maintenance capability for a six-ship lead package of F-22s	1	316
		HFBP3	Provides a dependent maintenance capability for a six-ship follow-on package of F-22s	1	166
	Munitions	HGBP1	Provides an independent munitions capability for a six-ship lead package of F-22s	1	90
		HGBP3	Provides a dependent munitions capability for a six-ship follow-on package of F-22s	1	32
192 FW	Operations	3FBP2	Provides additional operations personnel to support a follow-on six-ship package for contingencies or general war	1	28
		3FBP4	Provides additional operations personnel to support a follow on six-ship package for contingencies or general war	1	28
	Maintenance	HFBP2	Provides a dependent maintenance capability for a six-ship follow-on package of F-22s	1	226
		HFBP4	Provides a dependent maintenance capability for a six-ship follow-on package of F-22s	1	96
	Munitions	HGBP2	Provides a dependent munitions capability for a six-ship follow-on package of F-22s.	1	32
		HGBP4	Provides a dependent munitions capability for a six-ship follow-on package of F-22s.	1	16
Total				**13**	**1,150**

Each UTC describes both the number of authorized personnel (or equipment) and the Air Force specialty and skill level (or quantity and NSN for equipment). MEFPAK documents the UTC attributes, including the description, status, responsible organization, type, and transportation requirements. The Manpower Force Packaging component of MEFPAK describes the number of authorized personnel required to perform the mission and the Air Force specialty codes and skill level of each authorized individual, whereas the Logistics Force Packaging

System collects and describes materiel requirements for each UTC, such as the number and type of equipment.

Description of Capabilities as a Function of UTCs

Traditionally, war planners identify the groups of UTCs necessary to support a deployment request.[16] AFI 10-401 provides the policy for those war planners to develop and use UTCs in air expeditionary task force (AETF), functional areas, AETF support, and OPLANS and for developing and using force packages (FPs). Most of these, except for FPs, provide a broad organization of UTCs along functional areas. They require input from subject-matter experts to identify the correct mix of UTCs for a given deployment. An FP, however, is a collection of tailored UTCs grouped together to support various response options and is described in AFI 10-401 as providing a logical "grouping of records which facilitate planning, analysis, and monitoring." FPs are described as a method of packaging C2, operational mission, and expeditionary combat support forces for presentation to a combatant command. They are meant to provide packaged capability playbooks to support planning.

FPs provide a description of the capabilities needed in various planning scenarios. Through their organization and grouping of UTCs, FPs provide a link between personnel and equipment availability and capability. However, it is our understanding that FPs are not currently used for planning purposes because they are not yet fully developed.[17] Further, a complete listing of FPs and their UTC requirements is not readily available. Such a listing would provide a ready means to link personnel and equipment availability to needed capabilities for various planning scenarios and contingencies. If the grouping of UTCs into FPs is not available, groups of UTCs that have to work together to generate the needed capabilities must be identified. In the F-22 example, operation capability, maintenance capability, and munition capability are needed to provide air superiority capabilities.

Aviation packages used by ACC units do provide a limited listing of capability by type. As mentioned, a general aviation package is defined by the set of operation, maintenance, and munition UTCs required to generate lead and follow-on deployment packages (recall the example in Figure 2.3, Figure 2.4, and Table 3.3 for the F-22 squadrons). Table 3.4 presents a high-level description of the aviation packages and personnel UTC requirements for an F-22 squadron providing air superiority capability. Each UTC is described in terms of the number of personnel that it requires by AFSC.

---

[16] AFI 10-401, 2006.

[17] AFI 10-401, 2006, lists AF/A5XW as the office responsible for validating FPs annually and ensuring proper coordination for changes and deletions to UTCs in the FPs.

Table 3.4. Example of Personnel UTCs to Generate Aviation Packages

Element	Number of Required UTCs	Manpower Required (airmen)
Six-ship lead (6L)	3	498
6L + FW 1	6	784
6L + FW 1 + FW 2	9	1,010
6L + FW 1 + FW 2 + FW 3	12	1,150

An Analytical Model to Measure Unit Availability

The study team developed a UTC fulfillment model (Figure 3.1) to demonstrate how information about personnel, equipment, UTCs, and scenarios could be aggregated to assess readiness. The goal of the model is to bring together the UTC manning and equipment requirements (demand) for a given scenario and the training and availability of personnel and equipment of the wing and its units (supply). The model is designed to consider resources (personnel and equipment) from specific squadrons, specific Air Force components, or across squadrons and components when fulfilling UTC requirements.

Figure 3.1. Overview of UTC Fulfillment Modeling Approach

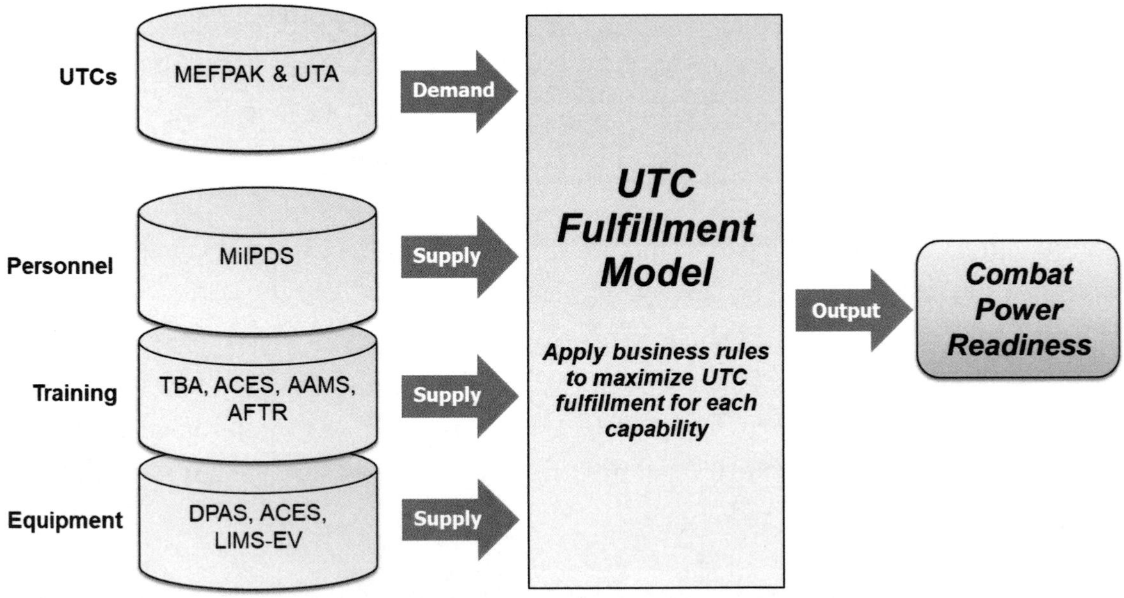

NOTES: UTA = UTC Availability; TBA = Training Business Area; ACES = Automated Civil Engineer System; AAMS = Automated Aircrew Management System; AFTR = Air Force Training Record; DPAS = Defense Planning Accountability System; LIMS-EV = Logistics Installation and Mission Support–Enterprise View. UTA is the data source that feeds ART. For more information, see AFI 10-244, 2012.

Ideally, information about the number and type of personnel, their training or qualification status, and the equipment of each squadron of a given wing would be available. Further, the data would be available at the individual level for personnel and at the part number for equipment. Although UTC requirements for personnel are expressed at the AFSC level (e.g., *X* personnel of *Y* AFSC), individual-level data are necessary to properly account for personnel who are both available and trained. Data at the part number level are necessary because the UTC requirements for equipment are expressed in terms of NSN.

The basic algorithm of the model works as follows:

- Step 1: Identify the capability being evaluated.
- Step 2: Select the UTCs needed to provide that capability.
- Step 3: List the manpower and equipment requirements for each of those UTCs.
- Step 4: Look across the wing for available and trained personnel (or equipment) who match the required AFSC and skill level (or NSN for equipment) or qualification level (e.g., four-ship flight lead) for each UTC.
- Step 5: Fill in each UTC with the available personnel and equipment as required, and note any deficiencies.
- Step 6: Repeat steps 1–5 for the next needed capability.

Business rules that reflect the goals and priorities of the wing, ACC, and the Air Force are needed to make the model work. The scenario for which readiness is being assessed drives the selection of the capability in step 1. For example, a wartime deployment capable of defeating aggression by a major power would require the availability of entire fighter squadrons, whereas preparedness to sustainably compete to deter aggression would require the availability of force packages (e.g., lead package). This means that, when assessing readiness for a wartime scenario, combat capability is defined by an entire squadron; however, for a competition scenario, combat capability may be defined only by lead packages. This business rule is implemented in the UTC fulfillment model by defining the order in which UTCs are filled. Figure 3.2 presents these two rules through the notional example of an F-22 wing (e.g., 1 FW and 192 FW at Langley AFB).

Figure 3.2. Business Rules That Capture UTC Fulfillment Priority for Wartime and Competition Scenarios

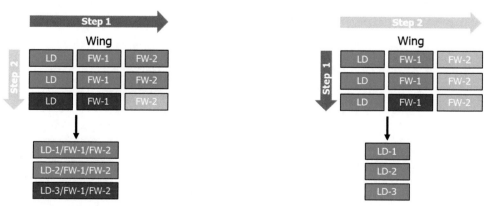

(a) Wartime scenario: generate full squadrons
NOTE: LD = lead.

(b) Competition scenario: generate lead packages

In a scenario requiring entire squadrons, the goal in fulfilling UTCs should be filling the UTCs that enable the generation of the lead and follow-on elements of the first squadron (step 1) and then the lead and follow-on elements of any subsequent deployments in any given wing (step 2). In the notional example shown in the left panel of Figure 3.2a, this means that two full squadrons are generated, and the remaining manpower is insufficient to generate a third squadron. In other cases, equipment could be the limiting factor.

In a scenario that requires only lead force packages, the goal in fulfilling UTCs is filling the UTCs that generate lead packages (step 1) and then (step 2) filling UTCs that generate the follow-on elements. Using the same manpower availability data, this means that three lead packages can be generated from the same wing (the right panel of Figure 3.2), with enough manpower availability to generate two follow-on elements.

The pool of resources available at the wing level to generate units must also be considered, since no individual squadron can generate capability by itself. Similarly, total force wings that rely on the availability and support of both active duty and reserve squadrons must consider resources in both types of units. Of relevance in total force wings will be assumptions about the level of volunteerism or activation of the reserve component units.

Example of Proposed Readiness Metric

To illustrate the utility of such a measure, consider the *personnel* requirements to fill the aviation UTCs (operation, maintenance, and munition personnel UTCs) of each wing across ACC. By using the personnel availability data of each airman in each ACC unit provided in MilPDS and the personnel requirements of each UTC (by AFSC and skill level) as described in the MEFPAK, we can assess the ability of each wing to generate the lead and follow-on combat capabilities. Figure 3.3 presents this assessment from a manpower perspective for a wartime scenario (i.e., the goal is to generate full deployable squadrons) and highlights the combat-power-generation capability for each mission design series (the right panel of Figure 3.3).

The entries in the table indicate the total shortfall in personnel to fill the UTCs that compose the lead and follow-on (FW 1, FW 2, FW 3) packages. The assessment of green, yellow, and red is arbitrary but shows how the databases can be used to calculate the number of deployable squadrons (complete lead + FW 1 + FW 2 + FW 3 packages, shown in the right panel of Figure 3.3). For example, a total of 12 squadrons can be considered available (two F-15E, four F-16, one F-22, two F-35, and three A-10 squadrons). This assessment assumes that all personnel in an AFSC at a given base are available to fill UTCs, regardless of whether they are in an active duty, ANG, or reserve unit. It is possible, and at times may be necessary, to limit the fulfillment of required UTCs to a specific component, base, or set of squadrons (e.g., ARC units might not be 100 percent available, especially for scenarios that do not merit a large reserve mobilization).

Figure 3.3. Combat-Power-Generation Capability of Manpower for *Wartime* (emphasizing entire 24-aircraft squadron employments, with lead and multiple follow-on packages)

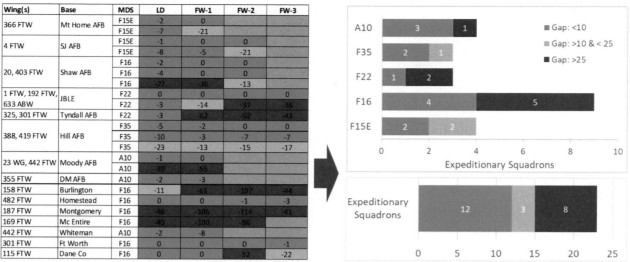

Wing(s)	Base	MDS	LD	FW-1	FW-2	FW-3
366 FTW	Mt Home AFB	F15E	-2	0		
		F15E	-7	-21		
4 FTW	SJ AFB	F15E	-1	0	0	
		F15E	-8	-5	-21	
20, 403 FTW	Shaw AFB	F16	-2	0	0	
		F16	-4	0	0	
		F16	-27	-36	-13	
1 FTW, 192 FTW, 633 ABW	JBLE	F22	0	0	0	0
		F22	-3	-14	-37	-36
325, 301 FTW	Tyndall AFB	F22	-3	-62	-52	-41
388, 419 FTW	Hill AFB	F35	-5	-2	0	0
		F35	-10	-3	-7	-7
		F35	-23	-13	-15	-17
23 WG, 442 FTW	Moody AFB	A10	-1	0		
		A10	-39	-55		
355 FTW	DM AFB	A10	-2	-3		
158 FTW	Burlington	F16	-11	-61	-107	-44
482 FTW	Homestead	F16	0	0	-1	-3
187 FTW	Montgomery	F16	-46	-106	-114	-41
169 FTW	Mc Entire	F16	-40	-100	-86	
442 FTW	Whiteman	A10	-2	-8		
301 FTW	Ft Worth	F16	0	0	0	-1
115 FTW	Dane Co	F16	0	0	-32	-22

NOTE: LD = lead; FTW = flying training wing ; DM = Davis-Monthan.

One can conduct a similar assessment for a competition scenario emphasizing smaller rotational packages (i.e., the goal is to generate a lead package). Figure 3.4 presents this assessment from a manpower perspective.

Figure 3.4. Combat-Power-Generation Capability of Manpower for *Competition* (emphasizing lead packages)

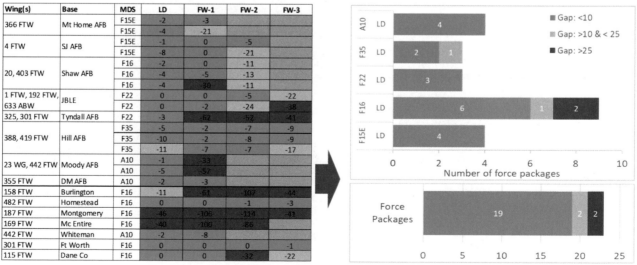

Wing(s)	Base	MDS	LD	FW-1	FW-2	FW-3
366 FTW	Mt Home AFB	F15E	-2	-3		
		F15E	-4	-21		
4 FTW	SJ AFB	F15E	-1	0	-5	
		F15E	-8	0	-21	
20, 403 FTW	Shaw AFB	F16	-2	0	-11	
		F16	-4	-5	-13	
		F16	-4	-30	-11	
1 FTW, 192 FTW, 633 ABW	JBLE	F22	0	0	-5	-22
		F22	0	-2	-24	-38
325, 301 FTW	Tyndall AFB	F22	-3	-62	-52	-41
388, 419 FTW	Hill AFB	F35	-5	-2	-7	-9
		F35	-10	-2	-8	-9
		F35	-11	-7	-7	-17
23 WG, 442 FTW	Moody AFB	A10	-1	-33		
		A10	-5	-57		
355 FTW	DM AFB	A10	-2	-3		
158 FTW	Burlington	F16	-11	-61	-107	-44
482 FTW	Homestead	F16	0	0	-1	-3
187 FTW	Montgomery	F16	-46	-106	-114	-41
169 FTW	Mc Entire	F16	-40	-100	-86	
442 FTW	Whiteman	A10	-2	-8		
301 FTW	Ft Worth	F16	0	0	0	-1
115 FTW	Dane Co	F16	0	0	-32	-22

NOTE: LD = lead; FTW = flying training wing ; DM = Davis-Monthan.

Because the goal in this scenario is to fill lead packages, the requirement is less demanding, and thus more packages can be generated (in this example, 19).

Although no information about training and equipment is included here, this representation shows how such a metric can provide a better picture of the number of deployable units.

Challenges in Implementing Proposed Readiness Measures

The current readiness metrics tracked in DRRS-S are too broad in some cases—and lack needed detail in others—to enable the proposed readiness representation and measurement. Similarly, the organization of UTCs is geared mostly toward managing UTCs rather than assessing their readiness for employment and whether the needed combinations to deliver capability are simultaneously ready. Although personnel and equipment data are available and tracked, they are disjointed and typically aggregated at too broad of a level to be of use here.

Personnel and Training Data

There is a wealth of information and data tracking the availability and training of personnel. However, the systems used to track these data are disjointed and do not readily allow the assessment of personnel availability and training in a readiness-relevant manner. For example, MilPDS reports the availability of personnel in each unit, whereas training data from Training Business Area report the training of personnel in the same unit. Current readiness analysis and reporting brings these two metrics together at the squadron level, independently assessing the availability and training of personnel (see the left panel of Figure 3.5). Although useful for identifying personnel availability and training shortfalls, current reporting fails to capture the full picture of personnel training and availability to support unit deployment because it is limited to the individual-unit level.

Figure 3.5. Tracking of Personnel and Training Data

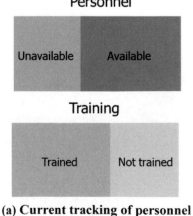

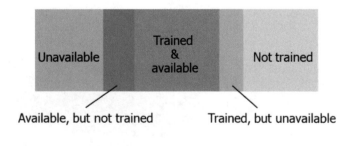

(a) Current tracking of personnel and training data

(b) Desired tracking of personnel and training data

Information that captures personnel availability and training in an integrated way is needed. Personnel availability and training data must be combined at the individual airman level to do so. DRRS-S contains data on personnel availability and training at the unit level. It might be possible to modify the database to categorize the status of individuals in a more operationally relevant way (see the right side of Figure 3.5).

Equipment Data

Data on the availability of equipment are currently available in DRRS-S. Further, an assessment of equipment that constitutes equipment UTCs is available in ART. In DRRS-S, the data are aggregated and reported at the unit level (a unit reports the percentage of its equipment that is on hand and available). Unfortunately, this is too coarse to enable the identification of available equipment to fill UTCs. Although a grouping of equipment into critical equipment packets is also available in DRRS-S, it is not enough for use in the approach proposed here. Equipment (and personnel) data reported in ART provide a better assessment of relevant groups of equipment (and personnel). However, because each unit reports in isolation, it is not possible to assess the ability to fill multiple UTCs. Developing and making available the UTC interdependencies in DRRS-S could leverage the existing data and enable the assessment of readiness of combat power, not just individual units or individual capabilities.

Summary

The chapter presented an approach to determining which combination of UTCs would be needed for a given scenario. Scenarios would be evaluated categorically across six dimensions, from simple to highly complex. Challenging scenarios, in which the Air Force would add a capability at a bare base, for example, would require more UTCs than increasing the capacity of an existing capability operating at an established main base. An example of employing F-22 packages was presented showing the personnel requirement for various packages. Given the requirements of a scenario and the required UTCs, the proposed allocation model will seek to fill as many sets as possible from the supply. The supply could be personnel and equipment sourced from a single wing or base or across multiple bases. We showed how the readiness evaluations may differ depending on the goal of the scenario (e.g., filling as many lead packages as possible versus filing as many complete 24 Primary Aerospace Vehicle Authorizations [PAA] packages as possible). Challenges to implementing this type of approach are data availability and format. One data issue is that the training and personnel data are independent. Another data issue is that the equipment data are incomplete and not at the level of detail required to fill UTCs.

4. Summary and Recommendations

Summary

Current force readiness and availability metrics have important shortfalls that limit their ability to inform decisionmakers about the readiness of the force and the personnel and equipment availability to meet scenario demands.

In this project, we developed a new approach to leverage and aggregate authoritative data sources that feed the current readiness reporting system to better measure the ability of ACC forces to meet scenario demands. Force availability would be defined by linking the status of personnel and equipment to UTC combinations, which in turn provide needed capabilities.

Recommendations

To fully implement the readiness assessment approach described here, the following important steps should be taken:

- **Track and make available personnel data at the individual level and equipment data at the part level.** Current DRRS-S data do not contain enough detail to assess the availability of specific AFSCs (for personnel) or NSNs (for equipment) needed to fill UTCs. Adding such data at the level of detail needed would enable the use of aggregation tools to dynamically fill the combinations of UTCs called for by various employment scenarios. The additional tracking of data should be implementable via a relatively modest tool that leverages data available within the service and need not require an expansive enterprise-wide, all-encompassing software solution.
- **Explicitly link personnel availability data to personnel training data.** Personnel availability and personnel training data at the individual level are needed to assess UTC readiness. At present, these are tracked separately, leaving important information and knowledge gaps.
- **Define or update descriptions and groupings of UTCs that enable their aggregation into relevant FPs.** Combinations of UTCs are needed to employ forces. An explicit listing of the UTCs needed to provide a given capability under a set of conditions is required to evaluate readiness in an operationally relevant manner. Such a listing could allow force structure and operation planners to accurately assess the supply of and demand for forces.[18]

---

[18] The guidelines for the definition and description of FPs seem to be an existing approach to accomplish this. However, we were not able to find a complete listing of FPs and their descriptions. If such a listing exists or is being developed, it would be a great step in assessing the readiness metric proposed here.

Along with a refinement of the scenario types and description of deployment conditions and requirements, it would be possible to measure and assess the readiness of combat power capability by utilizing a fulfillment model to generate the scenario-relevant UTCs, such as the one developed and used in this project.

Appendix A. Capability and Resource Readiness Reporting

In this appendix, we discuss capability and resource readiness reporting. Resource readiness is expressed by *category level* (C-level), which is defined by the Joint Staff and derived from quantitative criteria according to authoritative data sources. C-levels use a six-point scale to represent how well a unit is resourced. A rating of C-1 indicates the highest level of resourcing, whereas a rating of C-4 indicates the lowest level of resourcing. C-5 is reserved for units that are undergoing transition to a different mission and are not yet prepared to undertake the newly assigned mission, and C-6 indicates that a unit is not required to measure or report in a specified area. Table A.1 contains the specific definitions of the different levels from AFI 10-201.[19]

Table A.1. Definitions of Resource Readiness Assessment Scale

C-Level	Definition
1	Unit possesses required manpower, equipment, and training to perform all the missions for which it is organized or designed
2	Unit possesses required manpower, equipment, and training to perform most of the missions for which it is organized or designed
3	Unit possesses required manpower, equipment, and training to perform many, but not all, the missions for which it is organized or designed
4	Unit requires additional manpower, equipment, and training to perform the missions for which it is organized or designed
5	Unit is undergoing a service-directed resource action and is not prepared to perform the missions for which it is organized or designed
6	Unit is not required to measure or report in a specified area

SOURCE: AFI 10-201, 2016.

The C-level is based on separate ratings in four resource areas: personnel (P), equipment and supplies on hand (S), equipment condition (R), and training of assigned personnel (T). The resource readiness in each is also reported using the same six-point scale. Importantly, the worst of the individual areas determine the overall C-level. The ratings for the four areas are automated based on business rules applied to authoritative data sources that reflect metrics and criteria appropriate for each area. Currently, operational aviation units report against only personnel and training of assigned personnel, whereas the maintenance and supply units associated with the operational aviation units report in all four areas. Resource ratings are largely considered to be an objective assessment of unit readiness.

---

[19] AFI 10-201, 2016.

Capability readiness is a more subjective assessment based on the unit commander's evaluation of the unit's ability to conduct specified tasks across a set of OPLANs and named operations. The assessment should be based on observed performance, resource availability, military experience, and judgment. Capability readiness is reported as a yes (Y), qualified yes (Q), or no (N); units that do not report capability readiness are reported as not reporting (NR) (Table A.2).

Table A.2. Definitions of Capability Readiness Assessment Scale

Level	Assessment	Definition
Y	Yes, green	Unit can accomplish task to established standard(s) and condition(s)
Q	Qualified yes, yellow	Unit can accomplish all or most of the tasks to standard under most conditions
N	No, red	Unit is unable to accomplish the task to prescribed standard(s) and condition(s) at this time
NR	Not reporting	Unit does not report in a specific task

SOURCE: AFI 10-201, 2016.

Within capability readiness, units are rated against a set of METs. All METs fall within a unit's core mission set. Some subset of the METS is assessed in two other categories: OPLANs and named operations. Each of the METs is rated using the same yes, qualified yes, and no scale. The unit commander takes the aggregate scoring of each of the METs to report the overall capability rating for the reported readiness of ACC units.[20]

Both resource and capability assessments are done for the entire unit and reflect the overall readiness of the unit, without consideration of capabilities within the unit. For example, the 27th Fighter Squadron at Langley AFB reports its resource and capability readiness for the entire squadron, not based on its ability to generate a six-ship lead package or any combination of lead and follow-on packages.

---

[20] There is allowance for commander judgment in determining the overall unit capability rating. The guidance to commanders for this determination is not the "worst" MET but rather a formula described in AFI 10-201, 2016, that relies on the proportion of METs rated as yes or qualified yes.

Appendix B. Description of UTCs from MISCAPs and MEFPAK

This appendix provides a description of UTCs from MISCAPs and the MEFPAK.

Overview of MISCAP and the MEFPAK

The MEFPAK database's MISCAPs are a primary source of UTC data used for the analysis presented in the main report. Each UTC has a statement describing the UTC tasking and the relationship between the UTC and other packages. Our objective was to use the MISCAP descriptions to identify interdependencies among UTCs and use that information to group UTCs into capability packages. Figure B.1 presents a limited view of the interdependencies among the F-22 UTCs at Langley AFB.

Figure B.1. Network Representation of UTCs as Described in MISCAP for Select UTCs at 1 FW, 192 FW, and 633 ABW

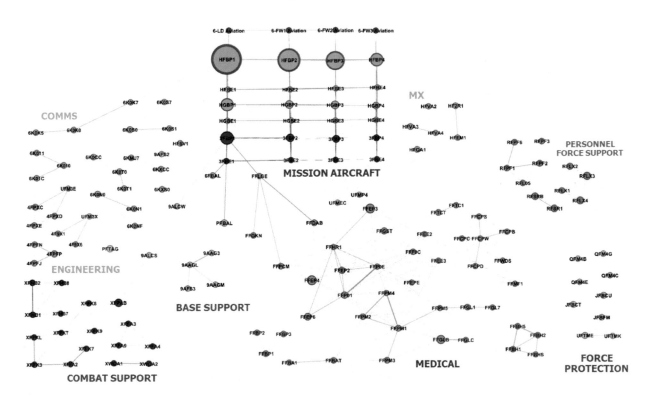

NOTE: LD = lead; COMMS = communications; MX = munitions.

We had planned to use text analysis to identify grouping of UTCs that define a capability. However, several limitations in the data and the MISCAP descriptions themselves made this impossible. The shortfalls and inconsistencies in the descriptions identified during the data-cleaning process offer an opportunity to improve the descriptions and enable their use as a source for describing UTC interdependencies. Because UTCs are a building block for force presentation, clearly and concisely describing the capabilities and dependencies of each UTC can greatly expand their utility. We discuss this below.

Problems with Text Characteristics

This MISCAP file is essential for understanding the interrelationships between UTC packages. That is difficult, given the problems with the data. Text descriptions of the mission statements lack standardized language, sometimes use shorthand notation, contain inaccurate language or typos within the text, and can have incomplete or obsolete descriptions. As a result, it is difficult to properly leverage these descriptions to piece together all the dependencies across UTCs.

Lack of Standardized Language and Use of Shorthand Notations

The language in the statements is not standardized, with multiple words or phrases used to convey the same types of relationships. Although the basic verbs *supports* and *requires* are most common (e.g., *UTC A supports UTC B* or *UTC B requires UTC C*), synonyms and phrases often do not present clear directionality in package relationships—that is, the difference between which packages *support* others and which packages *are supported by* others. These types of synonyms or phrases include *may be used in conjunction with, may be supplemented by* or *may supplement*, and *augments*. An algorithm that is parsing the text to identify UTC relationships and limits the ability to identify directionality between packages has problems with this type of language.

Additionally, shorthand notation referencing UTCs is prevalent and inconsistent. This includes the common use of slashes and dashes to refer to multiple UTC codes (e.g., UTC A–D or UTC A/B/C/D). For example, the MISCAP statement for UTC 3F3P4 reads as follows:

> SUPPORTS UTC'S 3F3E4 AND 3F3P1/2/3 IF AFRC AND 3F3P1/2 IF NGB [National Guard Bureau] AND IS SUPPORTED BY UTC'S HE3P4, HF3P4, AND HG3P4. PROVIDES PERSONNEL TO SUPPORT AN ADDITIONAL 3 PMAI [Primary Mission Aerospace Vehicle Inventory] TO PROVIDE A TOTAL OF 18 PMAI FOR TFI UNITS AT AN FOL [forward operating location]. MUST DEPLOY WITH UTC3F3P1/2/3 . . .

In the above case, UTCs 3F3P1, 3F3P2, and 3F3P3 are combined with a slash. However, this method to refer to multiple codes is not the only one used in the statements; another method is the use of dashes between the first and last UTC code. (In this case, the method would produce the text *3F3P1-3.*) Although understandable to a reader, these shorthand notations may pose problems for an algorithm attempting to identify UTC codes within the text, especially when applied inconsistently.

Inaccurate Language or Typos Within the Text

The language occasionally contains typos, limiting the ability to detect UTCs within the text and possibly causing some UTCs to be missed. The most common typo identified is the presumably accidental omission of a space between a UTC and the preceding or subsequent word. Using the aforementioned description of the UTC 3F3P4 package, for example, notice the lack of a space between the abbreviation *UTC* and the subsequent code (emphasis added):

> MUST DEPLOY WITH **UTC3F3P1/2/3**.

Additionally, the following sentence is included in the description for the UTC package 1SCG1 (emphasis added):

> TOTAL OF 3 OF THIS UTC AND 3 OF UTC **1SCG4ARE** REQUIRED FOR 24 HOUR/7 DAY CAPABILITY.

Although these types of mistakes are trivial for a reader to detect and interpret, they may limit the ability of algorithms to quantify relationships between packages.

Descriptions Are Incomplete or Obsolete

Finally, not all the text statements describing UTC packages include a comprehensive list of the other packages related to the package mission. At times, statements omit references to existing UTCs that are related to the package's mission, refer to obsolete packages, and refer to packages that lack their own mission statement. Some examples include the following:

- UTC packages XFFG1, XFFG2, XFFG3, XFFG4, and XFFG5 are related. Yet the text description for XFFG1 references only XFFG5. This is a problem; because the other packages are not explicitly referenced, extrapolation would be required to identify relationships between the XFFG1 and the other packages.
- UTC packages 1SCG1, 1SCG3, and 1SCG5 all refer in their text descriptions to UTC package 1SCG4. However, 1SCG4 is not listed as a UTC in the MISCAP file because it was phased out after fiscal year 2019 (and the references to it mention that it is considered obsolete). An algorithm that identifies the UTC code 1SCG4 but fails to search for phrases that identify obsolete codes may either skip the relationship entirely or impute a relationship that is outdated.
- Although packages 3FVE1, 3FVP1, and HEVF2 reference the package 3FVE2, the package 3FVE2 is not listed as a UTC in the MISCAP file at all. Similarly, packages HGVP2 and HFVP2 reference the package 3FVP2, which is not listed as a UTC in the MISCAP file at all. An algorithm may impute nonexistent packages or omit relationships that may be important.

Cleaning Procedure

Because of the aforementioned issues with the MISCAP descriptions, the ability to analyze and visualize relationships between packages required automated cleaning and manual adjustments.[21]

Data processing occurred in two distinct phases: text cleaning and relationship identification. The first phase initially created an object with a row for each word for each UTC code. Punctuation and words that did not contain a UTC code were removed. Subsequently, a variety of rules were applied to words containing but not strictly matching a UTC code. This includes rules to address shorthand notation (e.g., converting the aforementioned 3F3P1/2/3 into 3F3P1, 3F3P2, and 3F3P3) and correct typos (e.g., converting the aforementioned 1SCG4ARE into 1SCG4). The product was an object with two variables: a column for each package and a column containing all the UTC codes mentioned in the text description of the package. For example, if the text description of UTC A contained references to UTC B and UTC C, then the object would contain the rows shown in Table B.1.

Table B.1. Example Linkage Description of UTCs

Package	UTC Codes Mentioned in Package Description
A	B
A	C

However, these objects contain numerous pairs of packages that may be considered redundant. For packages with a clear hierarchy, it was beneficial remove these redundant entries. This second phase of processing addressed the cases in the object where there were multiple packages and where the first four characters of the five-character code were identical and the last character was a digit. The goal of the subsequent cleaning steps was to preserve only the pairs that were hierarchical neighbors. For example, consider packages with the UTC codes XXXX1, XXXX2, and XXXX3, where the object contains the pairs presented in Table B.2.

---

[21] The MISCAP file, with the UTC codes and related text descriptions of mission statements, was processed primarily using the R statistical programming language (version 3.5.0) within the integrated development environment RStudio (version 1.2.1335) using a 2017 MacBook Pro with macOS Mojave (version 10.14.6). Output from the R session was saved to a Microsoft Excel file, where it was manipulated manually in Excel (version 16.35) prior to completion.

Table B.2. Example UTC Description

Package	UTC Codes Mentioned in Package Description
XXXX1	*XXXX2*
XXXX1	*XXXX3*
XXXX2	*XXXX3*

The second pair (XXXX1 and XXXX3) is considered redundant given that an intermediate pair (XXXX2 and XXXX3) exists. The relationship identification phase of the text cleaning process removes these redundant pairs when an intermediate pair of packages exists within a hierarchy of packages sharing the same first four characters of their UTC code. Given that some codes are non-sequential and that MISCAP descriptions may contradict a presumed hierarchy, this step was manually completed using Microsoft Excel.

Afterwards, the adjusted cases were combined with the non-applicable cases within the R programming environment. The pairs of packages and the UTC codes mentioned in the package descriptions were merged with other variables, including the UTC DEPID codes, functional area, and UTC titles. Finally, various subsets of the data were created, including subsets that removed packages that did not have any UTC codes contained in their MISCAP descriptions and subsets of packages for specific bases (notably, JBLE). These subsets and the set of all data were output as a Microsoft Excel file for analysis and visualization.

Recommendations for Data and Analytical Improvement

The complex cleaning required to identify relationships between packages using MISCAP text descriptions underscores the limitations of the existing MISCAP file. We offer three recommendations to improve future analysis.

Standardize Package Relationships

Various verbs are used in the MISCAP descriptions to describe relationships between packages, including synonyms (e.g., *supplement* and *augment*) and ambiguous phrases (*may be supplemented by* or *may be used in conjunction with*). Clarity can be established by reducing the number of types of relationships between packages. For example, there may be four types of relationships (and their converses) between packages: packages supporting other packages, packages requiring other packages, packages supplementing other packages, and packages replacing other packages. Establishing a definitive set of relationships and removing ambiguity in relationships can enhance future analysis.

Use Standardized Language and Notation

After the establishing a set of relationships between packages, clear terminology should be used to reduce ambiguity. This may be as straightforward as using the same verb (e.g., *supports*) to describe every relationship in which a package supports another. Further, a standard notation should be defined when referring to multiple UTC codes (i.e., using a dash to reduce multiple codes into a single phrase—for example, combining 3F3P1, 3F3P2, and 3F3P3 into 3F3P1-3), or a standard should be established that all codes are referred to individually. Additionally, it may be beneficial to describe each type of relationship in its own sentence. This may remove ambiguity that occurs when sets of packages are described as having different or multiple relationships with a package, typically in a long sentence.

Ensure Completeness

As identified above, MISCAP descriptions may refer to UTC codes that do not appear in the data or that may omit references to relevant packages. Future analysis would benefit from a standard that ensures description completeness and identifies cases in which a reference may be obsolete.

Move Away from Sentence-Based Format to Describe UTC Interdependencies

Generally, deriving information for analysis from text, such as the MISCAP descriptions, poses challenges because of the nature of language processing. It may be beneficial to consider the creation of data separate from the sentence-based format of the MISCAP descriptions. Specifically, analysis could be improved with the use of a file that, for each package, lists all the other packages and their relationships to the initial package in a matrix-like format.

As an example of this proposal, consider a matrix in which each row is a package and each column denotes a type of relationship between packages. Cells contain the packages with a relationship to the row package. In this example, consider that the only relationships are *requires*, *supports*, *replaces*, and *supplements* and consider package UTC A, B, C, D, and E. If UTC A requires UTC B and supports UTC C, while UTC B can be replaced by UTC D and can be supplemented by UTC E, then the matrix is as shown in Table B.3.

Table B.3. Proposed Description and Categorization of UTC Interdependencies

UTC	Requires	Supports	Replaces	Supplements
A	B	C		
B				
C				
D			B	
E				B

Using an example from the existing MISCAP file, consider UTC 3F3P4. The description of this package includes the following:

SUPPORTS UTC'S 3F3E4 AND 3F3P1/2/3 IF AFRC AND 3F3P1/2 IF NGB
AND IS SUPPORTED BY UTC'S HE3P4, HF3P4, AND HG3P4.

From this excerpt, the matrix would be constructed as shown in Table B.4.

Table B.4. Example Description and Categorization of UTC Interdependencies

UTC	Requires	Supports	Replaces	Supplements
3F3P4	HE3P4, HF3P4, HG3P4	3F3E4, 3F3P1, 3F3P2, 3F3P3		

Creating a definitive source of package relationship data would enable accurate analysis while limiting the likelihood of omissions or incorrect imputations from a language-processing algorithm.

Bibliography

Air Force Instruction 10-201, *Force Readiness Reporting*, Washington, D.C.: U.S. Air Force, March 3, 2016. As of April 20, 2020:
https://static.e-publishing.af.mil/production/1/af_a3/publication/afi10-201/afi10-201.pdf

Air Force Instruction 10-244, *Reporting Status of Air and Space Expeditionary Forces*, Washington, D.C.: U.S. Air Force, June 15, 2012. As of July 14, 2020:
https://static.e-publishing.af.mil/production/1/af_a3/publication/afi10-244/afi10-244.pdf

Air Force Instruction 10-401, *Air Force Operations Planning and Execution*, Washington, D.C.: U.S. Air Force, December 7, 2006. As of April 20, 2020:
https://static.e-publishing.af.mil/production/1/af_a3_5/publication/afi10-401/afi10-401.pdf

Air Force Instruction 10-403, *Deployment Planning and Execution*, Washington, D.C.: U.S. Air Force, April 17, 2020. As of July 14, 2020:
https://static.e-publishing.af.mil/production/1/af_a4/publication/afi10-403/afi10-403.pdf

Air Force Instruction 10-420, *Combat Air Forces Aviation Scheduling*, Washington, D.C.: U.S. Air Force, October 6, 2017. As of April 20, 2020:
https://static.e-publishing.af.mil/production/1/af_a3/publication/afi10-420/afi10-420.pdf

Chairman of the Joint Chiefs of Staff Instruction 3401.02B, *Force Readiness Reporting*, Washington, D.C.: Joint Chiefs of Staff, May 31, 2011.

Chairman of the Joint Chiefs of Staff Manual 3150.24E, *Type Unit Characteristics Report (TUCHAREP)*, Washington, D.C.: Joint Chiefs of Staff, August 10, 2018. As of July 14, 2020:
https://www.jcs.mil/Portals/36/Documents/Library/Manuals/CJCSM%203150.24E.pdf?ver=2018-08-23-152328-577

DoD—*See* U.S. Department of Defense.

U.S. Department of Defense, *Summary of the 2018 National Defense Strategy of the United States of America: Sharpening the American Military's Competitive Edge*, Washington, D.C., January 2018.